by Whitney Leigh
illustrated by Mary Thelen

Requests for permission to make copies of any part of the work should be addressed to School Permissions and Copyrights, Harcourt, Inc., 6277 Sea Harbor Drive, Orlando, Florida 32887–6777. Fax: 407-345-2418.

Printed in China

ISBN 10: 0-15-358418-1
ISBN 13: 978-0-15-358418-3

Ordering Options
ISBN 10: 0-15-358356-8 (Grade K On-Level Collection)
ISBN 13: 978-0-15-358356-8 (Grade K On-Level Collection)
ISBN 10: 0-15-360668-1 (package of 5)
ISBN 13: 978-0-15-360668-7 (package of 5)

4 5 6 7 8 9 10 0940 15 14 13 12 11 10 09

Here is a quiz.
I have a box.
What do I have in it?

Is a pet in the box?
Yes, it is a pet.

Can it run a lot?
It can run a lot.

Can it nap in a bed?
Yes, it can nap in a bed.

Can it kiss you?
Yes, it can kiss me.

Do you have a dog?
Is that what it is?

Yes, I have a dog.
That is what it is.